WHAT'S SUSTAINABILITY?

By Sophie Washburne

Published in 2025 by
KidHaven Publishing, an Imprint of Greenhaven Publishing, LLC
2544 Clinton Street
Buffalo, NY 14224

Designer: Deanna Lepovich
Editor: Jennifer Lombardo

Photo credits: Cover (top) Peter Adams Photography/Shutterstock.com; cover (bottom) Marina Lohrbach/Shutterstock.com; p. 5 TarasM/Shutterstock.com; p. 7 Olga Danylenko/Shutterstock.com; p. 11 INTREEGUE Photography/Shutterstock.com; p. 13 Everett Collection/Shutterstock.com; p. 15 Snapshot freddy/Shutterstock.com; p. 17 (main) Janet Julie Vanatko/Shutterstock.com; p. 17 (inset) anna.spoka/Shutterstock.com; p. 21 JBOY/Shutterstock.com.

Cataloging-in-Publication Data

Names: Washburne, Sophie.
Title: What's sustainability? / Sophie Washburne.
Description: Buffalo, New York : KidHaven Publishing, 2025. | Series: What's the issue? | Includes glossary and index.
Identifiers: ISBN 9781534547827 (pbk.) | ISBN 9781534547834 (library bound) | ISBN 9781534547841 (ebook)
Subjects: LCSH: Environmentalism–Juvenile literature. | Sustainable development–Juvenile literature. | Renewable energy sources–Juvenile literature. | Climatic changes–Juvenile literature. | Energy conservation–Juvenile literature.
Classification: LCC GE195.5 W37 2025 | DDC 363.7–dc23

Printed in the United States of America

Some of the images in this book illustrate individuals who are models. The depictions do not imply actual situations or events.

CPSIA compliance information: Batch #CSKH25: For further information contact Greenhaven Publishing LLC at 1-844-317-7404.

CONTENTS

Earth's Resources

Over thousands of years, people have learned to use Earth's many **resources** to make life more comfortable. Many people around the world have things such as running water in their home and a nearby grocery store that sells food from different parts of the world.

Today, we know that it's important to care about how we get and use those resources. We need to make sure we save some resources for the people who come after us. We also need to make sure we're using resources in ways that are safe. All these things are part of sustainable living.

Facing the Facts

Climate change is one of the biggest problems Earth is facing as a result of unsustainable living.

Many people enjoy eating foods that don't grow where they live. Trucks, boats, and planes carry food around the world. Having more choices can make meals more enjoyable, but we also must think about the pollution this travel causes.

Thinking Ahead

Imagine that you have a tub of ice cream in your freezer. Everyone in your family can have as much as they want, so everyone eats three bowls for dessert each night. By the middle of the week, the ice cream is gone, and you can't get more until next week. If everyone had eaten only one bowl each night, the ice cream would have lasted the whole week.

In this example, your family wasn't eating the ice cream sustainably. Sustainability means meeting the needs people have right now but also thinking about the needs they'll have in the **future**.

Facing the Facts

Other words people use for "sustainable" are "green" and "eco-friendly." These words have small differences, but people often use them interchangeably, or in a way where they all seem to mean the same thing.

It might seem like some of our resources, such as trees, will never run out. However, this can happen if people cut down trees faster than new ones can grow.

Gone Forever

When a plant or animal disappears from Earth, we say it has gone extinct. When it's close to being extinct, we call it endangered. Living things can go extinct for many reasons, including changes in their **habitat** and overuse by humans.

Sometimes extinction happens naturally. However, scientists say that human actions are causing a lot of species, or kinds of living things, to go extinct unnaturally and quickly. They say people need to find ways to live more sustainably so we can save the plants and animals that are still living on Earth.

Facing the Facts

As of 2023, there are more than 105,000 animals on the list of endangered species.

This graph shows that there has been a big rise in the number of extinct bird species. These bird species and many other species of plants and animals have gone extinct because of human actions.

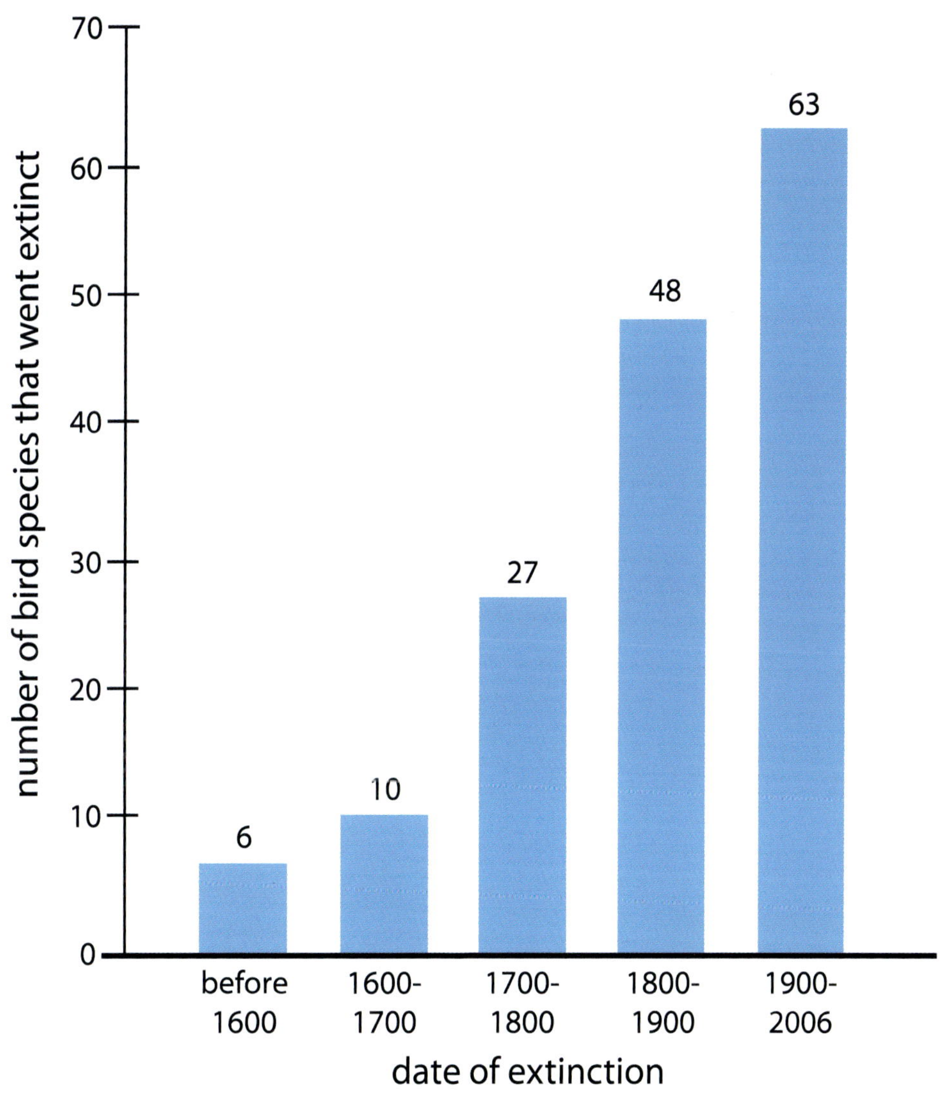

All Connected

To live sustainably, we need to think about the ways our actions affect everything around us. For example, some power plants burn coal to bring electricity to people's homes. The coal pollutes the air, which can make it hard to breathe. The pollution also kills plants and animals, which means there is less **biodiversity**.

When there isn't enough biodiversity, it's harder for nature to **adapt** to changes. Species become endangered or go extinct at a much faster rate. There may be shortages, or not enough of something to go around. For some people, a shortage is a little annoying. For others, it can be deadly.

Facing the Facts

A widespread shortage of food is called a famine. Famines can kill thousands of people.

Biodiversity is important because it creates backups. For example, if one species of wildflower is affected by a disease, there will still be other species left to give food to **pollinators**, such as honeybees.

The Dust Bowl

The Dust Bowl is one example of how important sustainability is in farming. In the 1920s, food prices were low. Farmers in states such as Kansas, Oklahoma, and New Mexico planted more crops so they could make more money. They dug up the grass of the Great Plains to create crop fields.

The grass the farmers dug up was holding the soil, or dirt, in place. When a harsh **drought** hit the area in the 1930s, high winds whipped up the soil and created many dust storms. Nothing could grow, and the dust storms covered everything for hundreds of miles. Many people had to move to other states.

Facing the Facts

Farms sometimes raise animals in unsustainable ways, such as cutting down trees to make room for more animals to feed. Also, cows release a gas called methane that traps heat in the **atmosphere**, which plays a part in climate change.

The Dust Bowl is considered one of the worst environmental, or natural, **disasters** in the history of the United States. This picture shows farm machines half-buried because of dust storms.

Clean Energy

Resources that will run out someday are called nonrenewable. Most of the things we use for **fuel**, such as coal, oil, and gas, are nonrenewable. These fuels also pollute the air and water.

Fuels that humans can easily get more of are called renewable. Most renewable resources are also clean sources of **energy**. This means they don't cause pollution. The sun and wind are two of the biggest clean, renewable energy sources. Using renewable fuel sources is more sustainable than using nonrenewable ones, so many places are working on changing where their energy comes from.

Facing the Facts

Water is a clean energy source. When it's used properly, water is also a renewable resource. However, many people waste water, which can cause droughts.

Solar panels and windmills take energy from the sun and wind. They turn that energy into a form that can power the things we use every day.

Too Many Things

Many companies make money by selling products, or goods. To keep making money, they need to keep making products. Then, the companies need to make people want to buy their products. Some companies make a lot of products very quickly and cheaply. The products break or rip, and then people throw them out and buy more.

It might not seem like a big deal to buy a new toy for $5, throw it out a year later when it breaks, and buy a new one. However, that creates a lot of waste. Making more products also uses a lot of energy and creates a lot of pollution.

Facing the Facts

"Fast fashion" is a name for cheap, poorly made clothes. Fast fashion is very unsustainable and causes a lot of problems for the environment. These items are also often made by people who must work long hours for low pay in bad conditions.

Cheap items are often not made to last for a long period of time. It's more sustainable to buy things that cost more and are made to last. However, not everyone can afford to do this.

Reduce, Reuse, Recycle

For many years, our economy—the way goods are made and sold—has been mostly linear. This means it goes in a straight line. Things are made, then sold, then used, then thrown away. However, a circular economy is more sustainable.

In a circular economy, goods are used and reused as much as possible. For example, instead of throwing away a piece of furniture when it starts to look old, a person might refurbish it, or make it look nice again. A circular economy creates less waste than a linear economy. Fewer things are made, and fewer things are thrown away.

Facing the Facts

Greenwashing is when a company tricks people into thinking it's using sustainable practices. This can be as simple as putting a label that says "eco-friendly" on a product that was not made sustainably.

The chart on the top shows a linear economy. The chart on the bottom shows a circular economy. In a circular economy, items are refurbished, repaired, or given away and reused to make them last longer. Items can also be recycled into something new when they can't be used anymore.

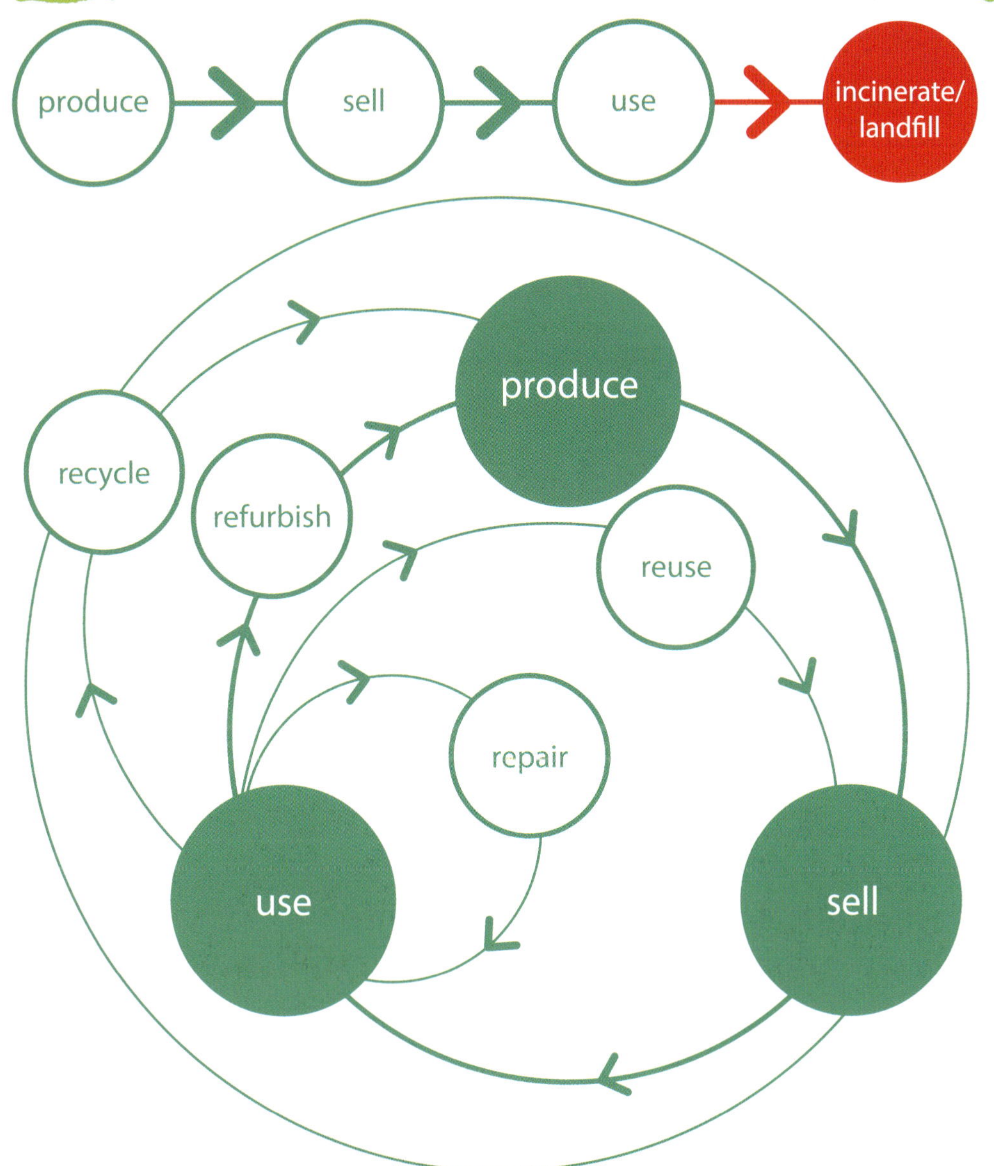

Do Your Part

It often feels hard to live sustainably. Cheap goods break quickly and need to be replaced often, but higher quality goods may cost more than people can afford. Large companies pollute the environment, and sometimes it feels like one person's actions aren't enough.

However, there are many things you can do to live more sustainably. Try to only buy products from sustainable companies. This sends a message to other companies that sustainability is important. Start small, and do what you can. Most of all, keep trying! Be kind to yourself, and remember that you're doing the best you can to help the planet.

Facing the Facts

Only 5 percent of the world's population lives in the United States, but it creates 40 percent of the world's waste.

WHAT CAN YOU DO?

Save water and energy by turning off sinks and lights when you aren't using them.

Buy things that will last a long time and reuse them.

Drink tap water instead of bottled water whenever possible, and eat more plant-based meals.

Shop at thrift stores more often.

Trade, sell, or give away items you don't want instead of throwing them away.

Do your best not to waste food.

Ride your bike or take the bus when you can.

Create a garden at home or at school with an adult's permission.

Learn about a company before you buy its products, and do your best to only buy from companies that use sustainable practices.

There are many small changes you can make that add up to a big difference.

GLOSSARY

adapt: To change over time.

atmosphere: The thick layer of gases that surrounds Earth.

biodiversity: The state of having a large number of different species of plants and animals.

climate: The weather over a long period of time.

disaster: Something that happens suddenly and causes much suffering or loss.

drought: A long period of time without rain.

energy: The power and ability to do work.

fuel: A source of energy.

future: Time that is to come.

habitat: The natural home of a plant or animal.

pollinator: An animal that transfers pollen from one plant to another.

resource: Something that can be used.

FOR MORE INFORMATION

WEBSITES

EPA: Recycle City

www3.epa.gov/recyclecity

Play three fun games to learn more about recycling, saving energy, and reducing waste.

***National Geographic Kids*: Sustainability**

www.natgeokids.com/uk/kids-club/cool-kids/general-kids-club/lets-make-a-change-sustainability

Learn more about what sustainability is and why it's important.

BOOKS

Craigie, Gregor. *Why Humans Build Up: The Rise of Towers, Temples and Skyscrapers*. Victoria, British Columbia: Orca Book Publishers, 2022.

Sarah, Rachel. *Climate Champions: 15 Women Fighting for Your Future*. Chicago, IL: Chicago Review Press, 2023.

Wheeler, Kate, and Trent Huntington. *Team Trash: A Time Traveler's Guide to Sustainability*. New York, NY: Holiday House, 2023.

Publisher's note to educators and parents: Our editors have carefully reviewed these websites to ensure that they are suitable for students. Many websites change frequently, however, and we cannot guarantee that a site's future contents will continue to meet our high standards of quality and educational value. Be advised that students should be closely supervised whenever they access the internet.

INDEX